UNTIL THE COWS COME HOME

WRITTEN AND ILLUSTRATED WITH HAND-COLORED PHOTOGRAPHS BY

PATRICIA MILLS

NORTH-SOUTH BOOKS / NEW YORK

Published in the United States by North-South Books Inc., New York.

Library of Congress Cataloging-in-Publication Data is available.
ISBN 1-55858-190-1 (trade binding)
ISBN 1-55858-191-X (library binding)

Designed by Marc Cheshire
1 3 5 7 9 10 8 6 4 2
Printed in Belgium

The photographs in this book are
platinum palladium prints that have
been hand colored with oil paint.

FOR JIM

Out on a country farm,

farmers gather

the harvest of hay.

Sheep graze side by side on the meadow.

The corncrib is nearly full

and wood stacked high for the cold days to come.

Out back, grapes for jam hide under tangled vines

as vegetables ripen in the garden.

Trousers and rag rugs, warmed by the sun, dry on the line.

Down near the road,

coneflowers bloom beneath the hemlocks,

and the garden yield sells to passersby.

Now and then,

a sweet wind moves through the fields,

and bluebirds pause in their flight.

In the late day, as the river quiets,

dark coal trains roll by,

and by,

until the cows come home,

and the world is still until another day.

A NOTE FROM THE AUTHOR

SEVERAL YEARS AGO my husband and I bought a cabin retreat in West Virginia, not far from where I lived as a child. Staying at the cabin, I rediscovered the gray-green Appalachian Mountains and the sturdy people who live in them.

Every time I returned to my harried life in New Jersey, the images of my days in the country lingered. So during each visit I began to write down my observations and to photograph the area around the cabin and the nearby hills and farms. I've spent countless hours exploring the country-side, trying to capture what this land means to me. The photographs in this book were taken over a period of more than two years.

I hope that in a quiet way this book will open children to the simple pleasures and the great beauty of unspoiled rural life.